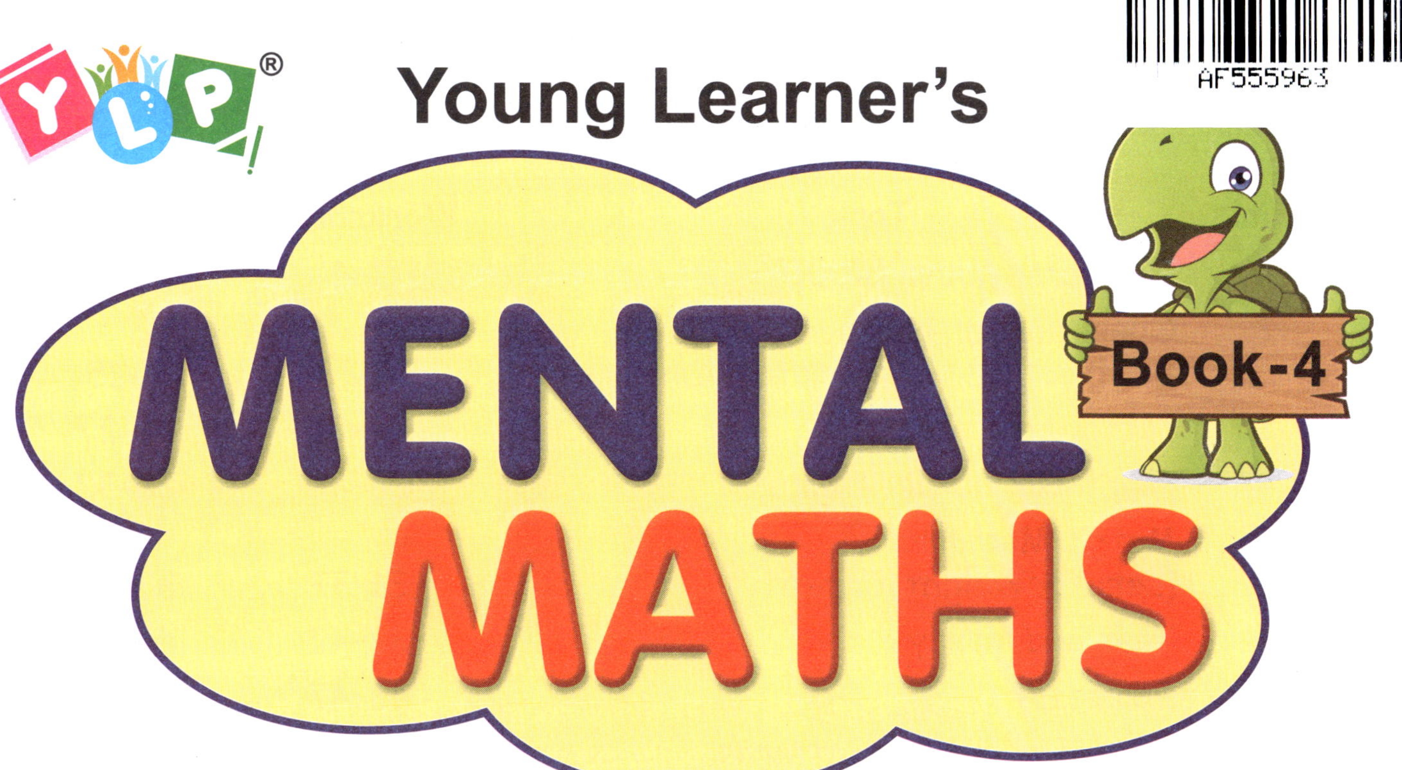

Rajesh Singh

CONTENTS

Published by:

YOUNG LEARNER PUBLICATIONS®

G-1A Rattan Jyoti, 18 Rajendra Place
New Delhi-110008 (INDIA)
Tel.: 011-25750801, 25820556, 25755559
Fax: 91-11-25764396
Email: goodwillpub@gmail.com
gph.ylp@gmail.com
gph.ylp@goodwillpublishinghouse.com
Website: www.goodwillpublishinghouse.com

Worksheet - 1

5 and 6 Digit Numbers, Place Value, Number Names

1. Write in ten thousands, thousands, hundreds, tens and ones.

27 653 ____

2. Write in hundred thousands, ten thousands, thousands, hundreds, tens and ones.

238 925 ____

3. Write the numeral for each of the following:

a) 6 ten thousands, 5 thousands, 4 hundreds, 3 tens and 5 ones __________

b) 8 hundred thousands, 3 ten thousands, 5 thousands, 3 hundreds, 2 tens and 8 ones __________________

4. Write the place value (PV) of each digit of the following numbers:

a) 95 127

PV of ___ is _________

PV of ___ is _________

PV of ___ is _________

PV of ___ is _________

PV of ___ is _________

b) 345 789

PV of ___ is _________

PV of ___ is _________

PV of ___ is _________

PV of ___ is _________

PV of ___ is _________

PV of ___ is _________

5. Write the following numbers in place value chart:

		H-Th	T-Th	Th	H	T	O
a)	69 758						
b)	753 546						

6. Write the number name for each of the following:

a) 12 983 ____

b) 237 459 ____

Worksheet - 2

Numerals, Successor and Predecessor

1. Write the numeral for each of the following:

a) Sixty-three thousand, eight hundred and seven ____________________

b) One hundred fifty thousand, two hundred and twenty ________________

2. Write the following in expanded form:

a) 71734 = ____

b) 345708 = ____

3. Write the numeral for each of the following:

a) 70000 + 3000 + 400 + 70 + 8 = ____________________________

b) 400000 + 50000 + 6000 + 40 + 8 = __________________________

4. Write the successor for each of the following:

a) 87314 ____________________ b) 652120 ____________________

c) Write next two numbers for 52439 ____________________________

d) Write next three numbers for 759309 ___________________________

5. Write the predecessor for each of the following:

a) 75428 ____________________ b) 552847 ____________________

c) Write two numbers that come just before 39802 ___________________

d) Write three numbers that come just before 478923 _________________

6. Fill in the blanks.

a) Write two numbers that come in between 24182, ____________________ ______________________, 24185

b) Write three numbers which come in between 630317, ________________ ______________________, 630321

c) Write four numbers that come in between 235192, __________________ ____________________, 235197

Worksheet - 3

Skip Counting, Comparison of Numbers, Largest and Smallest Numbers

1. Counting by two hundreds write three numerals from 25100.

____

2. Counting by five thousands write three numerals from 223000.

____

3. Counting by ten thousands write three numerals from 30001.

____

4. Counting by three hundred thousands write three numerals from 600000.

____

5. Tick the greater number: 241521 ☐ 241421 ☐
6. Tick the smaller number: 43269 ☐ 43134 ☐
7. Arrange in ascending order: 443329, 443257, 442727, 457420

____

8. Arrange in descending order: 32343, 44845, 34777, 44999

____

9. Make the smallest and largest 5 digit numbers using each of the digits 6, 3, 8, 1 and 4. Repetition is not allowed.

   Smallest: _______________ Largest: _______________

10. Make the smallest and largest 5 digit numbers using the digits 1, 2, 9, 5 and 7. Repetition is not allowed.

   Smallest: _______________ Largest: _______________

11. Make the smallest and largest 5 digit numbers using digits 4, 2, 0, 9, 5 and 8. Repetition is allowed.

   Smallest: _______________ Largest: _______________

12. Make the smallest and largest 5 digit numbers using digits 6, 3, 0, 1, 4 and 7. Repetition is not allowed.

   Smallest: _______________ Largest: _______________

13. Make the smallest and largest 6 digit numbers using each of the digits 2, 5, 8, 3, 7 and 6. Repetition is allowed.

   Smallest: _______________ Largest: _______________

Worksheet - 4

Largest and Smallest Numbers, Rounding-off Numbers

1. Make the smallest and largest 6 digit numbers using the digits 2, 0, 3, 7, 5 and 1. Repetition is not allowed.

Smallest: ______________ Largest: ______________

2. Make the smallest and largest 6 digit numbers using digits 2, 4, 6, 8, 3, 7 and 1. Repetition is allowed.

Smallest: ______________ Largest: ______________

3. Make the smallest and largest 6 digit numbers using digits 1, 0, 2, 4, 5, 6, and 7. Repetition is not allowed.

Smallest: ______________ Largest: ______________

4. Fill in the blanks.

a) Largest 5 digit palindrome* with three different digits is ______________.

b) Smallest 6 digit number more than 700000 is ______________.

c) Smallest number greater than 785781 is ______________.

d) Largest number less than 12454 and having all digits different is ________.

e) Smallest 5 digit number with all digits except the first and the last as 2's is ________________.

f) Largest 6 digit number without 3, 5, 7 and 9 and all digits different is ______.

g) Smallest 5 digit number with all digits different and in increasing order is ________________.

h) Largest 6 digit number without 0, 1 and 2 and all digits different is _________.

i) Smallest number between 31052 and 34359 having all digits same is ________________.

5. Round the following numbers to nearest tens:

a) 37425 ________________ b) 129031 ________________

6. Round the following numbers to nearest hundreds:

a) 47651 ________________ b) 439021 ________________

7. Round the following numbers to nearest thousands:

a) 55821 ________________ b) 871484 ________________

8. Round the following numbers to nearest ten-thousands:

a) 62464 ________________ b) 978697 ________________

9. Round the following numbers to nearest hundred thousands:

a) 726287 ________________ b) 878961 ________________

*Palindrome: A number that reads the same backwards as forwards, e.g. 121.

Unit-2
Operations on Numbers

Worksheet - 5

Addition Without Carrying and With Carrying

Solve the following:

1.

	2	4	7	5	3
+	4	3	1	2	4

2.

	6	7	2	5	3	4
+	2	2	5	4	3	3

3.

	1	3	4	5	7
	2	4	2	3	2
+	5	0	1	1	0

4.

	4	4	0	4	3	3
	2	2	6	2	5	2
+	3	2	2	3	1	4

5.

	4	6	7	7
+	8	5	8	2

6.

	2	3	6	7	2
+	8	7	8	7	8

7.

	2	7	8	3	6
	3	4	7	8	9
+	1	6	4	3	2

8.

	2	4	9	8	9	9
	3	7	7	7	4	8
+	2	9	3	5	6	7

9.

6	4	0	0	6	4
	1	0	9	1	9
+				2	0

10.

	1	4	0	1	9
	1	4	0	1	9
+	9	4	0	4	0

Worksheet - 6

Addition Stories

1. Population of a town was 357,451 two years ago. It has increased by 26,782 since then. What is the population of this town now?

2. In an election Mr. John got 50,481 votes, and Ms. Sophia got 63,490 votes. 2,320 voters did not turn up. How many voters in total voted for Mr. John and Ms. Sophia?

3. A website received 145,223 hits in first month and 145,123 hits in second month. In all how many hits did the website get in two months?

4. In a godown there are 235,364 sacks of rice, 26,657 sacks of millets and 4,582 sacks of lentils. What is the total number of sacks stored in the godown?

5. A factory produced 61,216 items in May and 65,421 items in June. How many items did it produce in two months?

6. A dealer sold 20,165 bicycles in the first year. In the second year he sold 31,011 bicycles and in the third year he sold 39,119 bicycles. How many bicycles did he sell in three years?

Worksheet - 7

Addition Facts, Missing Digits

Solve the following:

1. ________ + 45140 = 45140 + 24324

2. 56562 + ________ = 56646 + ________

3. [23074 + 74072] + 16041 = ________ + [74072 + 16041]

4. [74214 + 83985] + 98451 = ________ + [________ + 98451]

5. [________ + ________] + 37412 = 42546 + [54284 + ________]

6. 62352 + 0 = ________

7. 29252 + ________ = 29252

8. ________ + 0 = 46679

9. 0 + 74757 = ________

10. ________ + 88435 = 88435

11. 0 + ________ = 33169

12. **Find the missing digits.**

a

		3		2	1
+	3		4		6
	8	4	9	6	

b

	6		3		3	
+		4		3		4
	9	9	7	5	5	9

c

	1	*1*	*1*		
	5		8		5
+		4		9	
	9	2	5	8	7

d

	1	*1*	*1*	*1*	*1*	
		2	3		7	2
+	2	9		8	7	
	6		1	5		1

Worksheet - 8

Estimation by Rounding

Solve the following:

1. Estimate by rounding to nearest tens.

   24712 + 48628 ≈ ____________ + ____________ = ____________

2. A car dealer sold 18,527 cars in the first year and 16,314 cars in the second year. Estimate the number of cars sold by the dealer in the two years by rounding to nearest tens. ____________ + ____________ = ____________

3. Estimate by rounding to nearest hundreds.

   56235 + 41455 ≈ ____________ + ____________ = ____________

4. A machine produced 12,785 items in first shift and 13,432 items in second shift. Estimate the total number of items produced in two shifts by rounding to nearest hundreds. ____________ + ____________ = ____________

5. Estimate by rounding to nearest thousands.

   233597 + 107156 ≈ ____________ + ____________ = ____________

6. A trade fair in the city was visited by 43,269 people in the first week and 34,189 people in the second week. Estimate the number of people who visited the trade fair in the two weeks by rounding to nearest thousands.

   ____________ + ____________ = ____________

7. Estimate the sum by rounding to nearest ten thousands.

   124306 + 111898 ≈ ____________ + ____________ = ____________

8. Population of a town is 137,421. Population of a neighbouring town is 225,289. Find the total population of the two towns by rounding to nearest ten thousands.

   ____________ + ____________ = ____________

9. Estimate the sum by rounding to nearest hundred thousands.

   630921 + 257823 ≈ ____________ + ____________ = ____________

10. A car costs ₹ 283,225. Another car costs ₹ 436,410. Estimate the total cost of the two cars by rounding to nearest hundred thousands.

    ____________ + ____________ = ____________

Worksheet - 9

Subtraction Without Borrowing and With Borrowing

Solve the following:

1.

	5	7	8	5	5
–	2	3	7	2	4

2.

	9	6	8	7	6
–	3	4	6	3	2

3.

	4	0	5	5	7	3
–	1	0	2	3	5	2

4.

	7	8	4	5	4	8
–	2	2	2	3	2	1

5.

	8	4	5	3	6
–	7	6	6	3	8

6.

	6	3	8	6	3
–	3	7	6	8	9

7.

	7	4	6	2	1	3
–	2	3	8	7	7	7

8.

	9	6	0	6	2	3
–	2	2	4	7	8	4

9.

	9	4	0	0	1
–	1	9	9	1	0

10.

	9	0	1	0	0	1
–	8	7	1	5	5	5

Worksheet - 10

Subtraction Stories

1. In an election Ms. Nadia got 86,417 votes, and Mr. Simon got 89,541 votes. What was the victory margin?

2. A library has 147,390 books. Of this 66,510 books are in English. How many books are in other languages?

3. Two years ago population of a town was 534,700. It has increased to 614,630 since then. What is the increase in the population during the last two years?

4. 423,284 bags of corn were stored in a granary. Of this 324,900 were sold. How many bags of corn are left in the granary?

5. The total sale of a store during the month was ₹ 674,587 and expenditure was ₹ 374,908. What profit did the store make during the month?

6. A warehouse has 293,213 cement bags. Out of this 109,700 bags were despatched on the first day and 81,371 bags on the second day. What is the balance in the warehouse now?

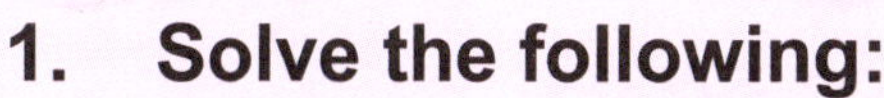

Worksheet - 11

Addition and Subtraction Together
Subtraction Facts, Missing Digits

1. Solve the following:

a) 53474 – 42418 + 11782

____

____

b) A factory produced 56,870 items in the first year and 42,720 items in the second year. The total sale during the two years was 90,875 items. How many items were in stock at the end of two years?

____

____

2. Fill in the blanks.

a) 34140 – 34140 = ______ b) 562376 – ______ = 0

c) ______ – 49214 = 0 d) 288352 – 0 = ______

e) 55219 – ______ = 55219 f) ______ – 0 = 867106

3. Check the following subtraction by addition:

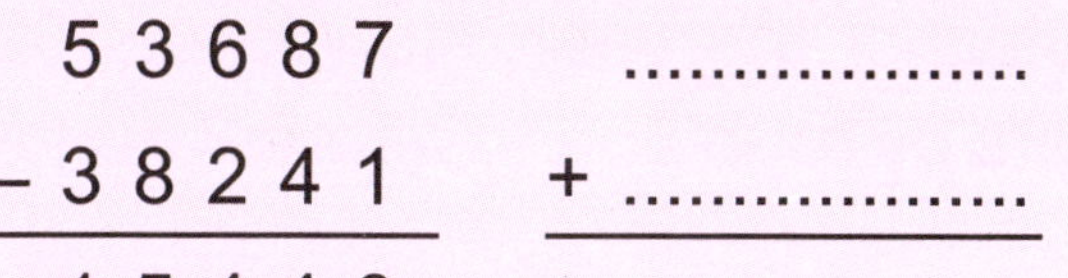

```
  5 3 6 8 7        ..................
– 3 8 2 4 1      + ..................
-----------      --------------------
  1 5 4 4 6        __________________
```

Subtraction is __________. (correct/incorrect)

4. Check the following addition by subtraction:

```
  3 6 8 5 2        ..................
+ 2 0 4 7 1      – ..................
-----------      --------------------
  5 7 3 2 3        __________________
```

Addition is __________. (correct/incorrect)

5. Find the missing digits.

a)

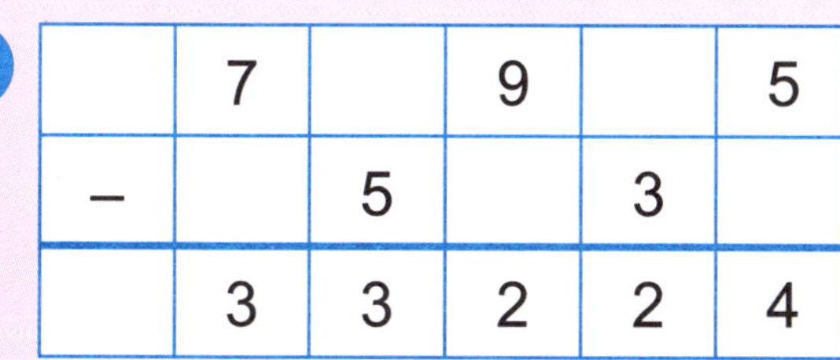

	7		9		5
–		5		3	
	3	3	2	2	4

b)

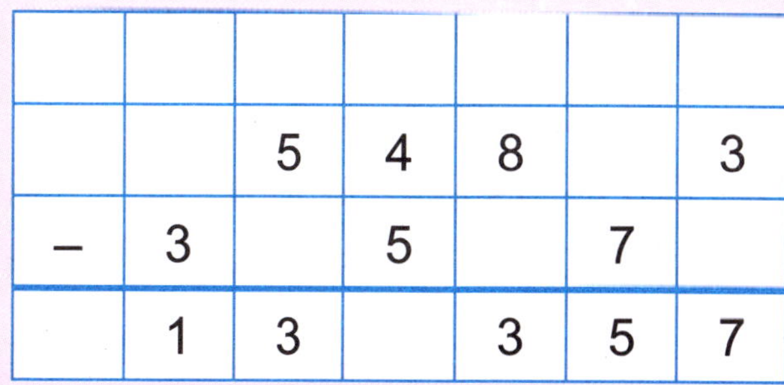

		5	4	8		3
–	3		5		7	
	1	3		3	5	7

Worksheet - 12

Estimating Difference, Estimation Stories

Solve the following:

1. Estimate the difference by rounding to nearest ten thousands.

73252 – 25254 ≈ ______________ – ______________ = ______________

2. A company manufactured 57,825 clocks in the first year and 68,254 clocks in the second year. Estimate how many more clocks were manufactured in the second year than the first year by rounding to nearest ten thousands.

______________ – ______________ = ______________

3. Estimate the difference by rounding to nearest hundred thousands.

765144 – 532582 ≈ ______________ – ______________ = ______________

4. In an election Mr. Yale got 736,874 votes. His nearest rival got 397,247 votes. Estimate the victory margin of Mr. Yale by rounding to nearest hundred thousands. ______________ – ______________ = ______________

5. Estimate the difference by rounding to nearest hundreds.

55392 – 37650 ≈ ______________ – ______________ = ______________

6. An online club had 29,840 members at start of the year. Now it has 32,270 members. Estimate the increase in members by rounding to nearest hundreds.

______________ – ______________ = ______________

7. Estimate the difference by rounding to nearest thousands.

567544 – 231406 ≈ ______________ – ______________ = ______________

8. A fun park was visited by 457,214 visitors in a month. Of these 238,741 visitors visited the fun park on working days. Estimate the number of visitors on holidays by rounding to nearest thousands.

______________ – ______________ = ______________

9. Estimate the difference by rounding to nearest tens.

76212 – 56539 ≈ ______________ – ______________ = ______________

10. Ten years ago population of a city was 119,727. Now its population is 290,258. Estimate the increase in population by rounding to nearest tens.

______________ – ______________ = ______________

Worksheet - 13

Multiplication Grid (Tables 11 to 20)

1. Fill in the boxes to complete the multiplication grid.

×	11	12	13	14	15	16	17	18	19	20
1										
2										
3										
4										
5										
6										
7										
8										
9										
10										

2. Each box has 8 soap cakes. How many soap cakes are there in 16 boxes?

   Number of soap cakes = ______ × ______ = ______

3. Each pouch has 7 pencils. How many pencils are there in 17 pouches?

   Number of pencils = ______ × ______ = ______

4. Each dress needs 9 buttons. How many buttons are needed for 18 dresses?

   Number of buttons = ______ × ______ = ______

5. Each pack has 10 toys. How many toys are there in 19 packets?

   Number of packets = ______ × ______ = ______

6. Each packet has 6 pens. How many pens are there in 20 packets?

   Number of pens = ______ × ______ = ______

Worksheet - 14

Vertical Multiplication (11 to 19)

Solve the following:

1

	5	7	8	9
		×	1	1

2

	6	4	6	2
		×	1	2

3

	5	5	3	1
		×	1	3

4

	4	9	8	7
		×	1	4

5

	4	5	7	6	9
			×	1	5

6

	5	4	9	7	3
			×	1	6

7

	3	5	8	4	3
			×	1	7

8

	4	5	3	1	7
			×	1	8

9

	3	5	8	4	3
			×	1	9

Worksheet - 15

Multiplication by Numbers Greater Than 20

Solve the following:

1

		3	5	8	6
			×	2	5

2

		7	3	7	2
			×	4	2

3

	1	3	0	8	7
			×	5	3

4

			5	0	4
		×	2	3	1

5

			5	6	7
		×	6	4	2

6

		1	4	0	3
		×	1	2	3

7

		1	6	0	8
		×	6	0	8

8

			4	9	4
		×	7	6	2

9

		2	4	2	9
		×	2	4	9

Worksheet - 16

Multiplication Stories

1. A lorry can carry 14,500 bricks in one round. How many bricks can it transport in 16 rounds?

____

____

2. Each bunch has 14 keys. How many keys will 2,016 such bunches have?

____

____

3. Each carton has 2,350 copies of a book. How many books will 37 such cartons have?

____

____

4. Each truck can carry 2,460 bags. How many bags can 135 trucks carry?

____

____

5. Each packet has 3,075 items. How many items will 18 such packets have?

____

____

6. A factory manufactures 5,400 pairs of shoes every month. How many pairs of shoes will the factory manufacture in 11 months?

____

____

Worksheet - 17

Multiplying by Multiples of 10

1. Fill in the blanks.

a) 1349 × 10 = ________________ b) 2715 × 100 = ________________

c) 49 × 1000 = ________________ d) 52 × 10000 = ________________

e) 5 × 100000 = ________________ f) 9 × 100000 = ________________

g) A carton has 10 items. How many items will 3,425 cartons have? ____________________________________

h) A packet has 100 erasers. How many erasers will 4,523 packets have? ____________________________________

i) A train can carry 1,000 passengers in a trip. How many passengers will it carry in 113 trips? ____________________________________

j) In each consignment 10,000 items are sent. How many items will be sent in 14 consignments? ____________________________________

k) A booklet has 8 pages. How many pages will 100,000 copies of the booklet have? ____________________________________

2. Solve the following:

a) 807 × 30 = ________________ b) 500 × 618 = ________________

c) 18 × 3000 = ________________ d) 13 × 30000 = ________________

e) A crate has 40 apples. How many apples will 1,255 crates have? ____________________________________

f) A ferry can transport 300 passengers in a trip. How many passengers will it transport in 213 trips? ____________________________________

g) A lorry can carry 2,000 bags of cement in one round. How many bags will it carry in 23 rounds?____________________________________

h) A carton has 20,000 items. How many items will 4 such cartons have? ____________________________________

Worksheet - 18

Multiplying Two Multiples of 10, Multiplication Facts, Estimating Product

1. Fill in the blanks.

a) 80 × 540 = ____________ b) 700 × 90 = ____________

c) 300 × 200 = ____________ d) 4000 × 20 = ____________

e) A book has 140 pages. How many pages will 90 copies of this book have?

____

f) A packet has 500 candles. How many candles will 190 packets have?

____

g) A carton has 3,000 items. How many items will 320 cartons have?

____

h) A book has 400 pages. How many pages will 2,000 copies of the book have?

____

i) 12563 × ________ = 28563 × 12563

j) [________ × 314151] × 807556 = 14349 × [314151 × 807556]

k) [50241 × 12874] × 25214 = ________ × [________ × 25214]

l) [________ × 57465] × 42140 = 20024 × [________ × ________]

m) 353374 × ________ = 353374 n) ________ × 1 = 47557

o) 132114 × 0 = ________ p) 44362 × ________ = 0

2. Estimate the product by rounding to nearest tens.

a) 417 × 67 ≈ ________ × ________ = ________

b) 5446 × 42 ≈ ________ × ________ = ________

c) A packet has 52 pens. Estimate the number of pens in 265 such packets by rounding to nearest tens.

________ × ________ ≈ ________ × ________ = ________

3. Estimate the product by rounding to nearest hundreds.

a) 114 × 92 ≈ ________ × ________ = ________

b) A bag has 125 letters. Estimate the number of letters in 414 such bags by rounding to nearest hundreds.

________ × ________ ≈ ________ × ________ = ________

Worksheet - 19

Long Division (Without Remainder)

Solve the following:

1. 86976 ÷ 8 Q = ______________	2. 86272 ÷ 16 Q = ______________
3. 829530 ÷ 18 Q = ______________	4. 644760 ÷ 45 Q = ______________

Worksheet - 20

Long Division (With Remainder)

Solve the following:

1. 86127 ÷ 9 Q = ___________. R = ___________.	2. 82280 ÷ 13 Q = ___________. R = ___________.
3. 230450 ÷ 42 Q = ___________. R = ___________.	4. 101233 ÷ 105 Q = ___________. R = ___________.

Worksheet - 21

Division Stories (Without Remainder)

Solve the following:

1 7 items are packed in 1 packet. How many packets are required to pack 21,763 items? ____________ packets.	2 One dozen eggs can be packed in a tray. How many trays are needed to pack 16,308 eggs? ____________ trays.
3 21,360 items are to be packed in packets of 20 items each. How many such packets can be packed? ____________ packets.	4 45,650 soap cakes are to be packed into packets of 25 soap cakes each. How many packets can be made?  ____________ packets.

Worksheet - 22

Division Stories (With Remainder)

Solve the following:

1. A dress requires 11 buttons. How many dresses can be buttoned using 34,415 buttons? How many buttons are left?

   ____ dresses. ____ buttons left.

2. 54,723 items have to be divided into 13 equal heaps. How many heaps can be made? How many items will be left behind?

   ____ heaps. ____ items left behind.

3. 234,550 items are to be packed in packets of 30 items each. How many such packets can be made? How many items are left unpacked?

   ____ packets. ____ items left unpacked.

4. At a bakery 125,825 biscuits have to be packed into tins of 300 biscuits each. How many tins can be filled? How many biscuits are left unpacked?

   ____ tins. ____ unpacked biscuits.

Worksheet - 23

Division by Multiples of 10 and Division Facts

1. Divide the following (without actual division):

a) 15870 ÷ 10 = ______ b) 697640 ÷ 10 = ______

c) 24900 ÷ 100 = ______ d) 723800 ÷ 100 = ______

e) 32000 ÷ 1000 = ______ f) 812000 ÷ 1000 = ______

g) 40000 ÷ 10000 = ______ h) 990000 ÷ 10000 = ______

i) 500000 ÷ 100000 = ______ j) 800000 ÷ 10000 = ______

2. Without actual division find the remainder when each of the following is divided by 10. [Hint: Ones digit if not zero, leaves a remainder.]

a) 56786, Remainder = ______ b) 615218, Remainder = ______

3. Fill in the blanks.

a) 23043 ÷ 23043 = ______ b) 133413 ÷ ______ = 1

c) ______ ÷ 52545 = 1 d) 624511 ÷ 1 = ______

e) 24467 ÷ ______ = 24467 f) ______ ÷ 1 = 157437

g) 0 ÷ 36347 = ______ h) ______ ÷ 564712 = 0

i) Given that, 62730 ÷ 17 = 3690. Then __________ = __________ × __________

j) If on dividing 11012 by 21 we get 524 as quotient and 8 as remainder then __________ = __________ × __________ + __________

[Hint: dividend = divisor × quotient + remainder]

Worksheet - 24

Multiples, Even and Odd Numbers

1. **Fill in the blanks.**

a) Write the first five multiples of 18 ____________________

b) Write the next five multiples after 16, 32, 48, 64, ____________________

c) Write multiples of 20 which are greater than 40 but less than 100.

____

d) Smallest multiple of 12 which is greater than 95 is ____________________.

e) Largest two digit multiple of 11 is ____________________.

f) Smallest three digit multiple of 19 is ____________________.

g) A multiple of 20 nearest to 142 is ____________________.

h) Next two even numbers after 14346 are ____________________.

i) Next two odd numbers after 15433 are ____________________.

j) Write the two even numbers just before 12370 ____________________

k) Write the two odd numbers just before 52403 ____________________

l) Largest two digit even number is ____________________.

m) Smallest three digit odd number is ____________________.

n) Smallest four digit even number is ____________________.

o) Largest five digit even number is ____________________.

p) Largest six digit odd number with all digits same is ____________________.

2. **Write True or False.**

a) 144 a multiple of 16 ________ b) 136 a multiple of 14 ________

c) 280 a multiple of 20 ________ d) 135 a multiple of 15 ________

3. **Tick the even numbers, cross out the odd numbers.**

1234, 5681, 3403, 27872, 7187, 48418, 6729, 5686, 85955, 901294

Worksheet - 25

Even and Odd Numbers, Factors

1. **Fill in the blanks.**

   a) Smallest five digit even number with all digits different is __________.

   b) Largest six digit odd number with all digits different is ___________.

   c) Smallest six digit even number with all digits different and even is _________.

2. **Write all factors for each of the following:**

   a) 10 ________________________ b) 32 ________________________

   c) 36 ________________________ d) 23 ________________________

3. **Are there some numbers which don't have factors other than 1 and themselves? (Y/N)**

   ____

4. **Give all odd factors for each of the following:**

   a) 15 ________________________ b) 37 ________________________

   c) 28 ________________________ d) 40 ________________________

5. **Can a number which is even have an odd factor? (Y/N)** ________________

6. **Give all even factors for each of the following:**

   a) 44 ________________________ b) 50 ________________________

   c) 81 ________________________ d) 99 ________________________

7. **Can a number which is odd have an even factor? (Y/N)** ________________

8. **Give factors having two digits for each of the following:**

   a) 144 ________________________ b) 200 ________________________

9. **Can a 3 digit number have a factor which is also a 3 digit number? (Y/N) If yes, give an example.** ____

Worksheet - 26

Fractional Names, Numerator and Denominator, Comparing Fractions

1. Write the fraction for each of the following:

a) Two-third: ______________ b) Three-fourth: ______________

2. Write the fractional name for each of the following:

a) $\frac{2}{5}$: ______________________ b) $\frac{3}{8}$: ______________________

3. Fill in the blanks.

a) For $\frac{12}{19}$; Numerator =_________, Denominator = _________.

b) Fraction with numerator 3 and denominator 7 is ______________.

c) Fraction with denominator 10 and numerator 3 is ______________.

d) Fraction with numerator 4 and denominator 9 more than the numerator is ______________.

e) Fraction with denominator 7 and numerator 5 less than the denominator is ______________.

f) Fraction with numerator as the largest one digit even number and denominator as the smallest two digit odd number is ______________.

g) If two fractions have the same denominator, then the fraction having the smaller numerator represents the ______________ fractional number. (smaller/greater)

h) If two fractions have the same numerator, then the fraction having the ______________ denominator represents the smaller fractional number. (smaller/greater)

4. Compare and write the correct sign (> or <) in each of the following:

a) $\frac{4}{7}$ ☐ $\frac{5}{7}$ b) $\frac{5}{9}$ ☐ $\frac{2}{9}$ c) $\frac{1}{4}$ ☐ $\frac{1}{2}$

5. Arrange the following in ascending order of value:

$\frac{5}{8}, \frac{1}{8}, \frac{7}{8}$: ____

6. Arrange the following in descending order of value:

$\frac{5}{8}, \frac{5}{11}, \frac{5}{7}$: ____

Worksheet - 27

Equivalent Fractions

Fill in the blanks.

1. Equivalent fractions of a fraction are formed by __________ or __________ the numerator and the denominator by the same number (other than 1 and zero).

2. Next three equivalent fractions for $\frac{2}{7}, \frac{4}{14}, \frac{6}{21}$ are ______________________.

3. First three equivalent fractions of $\frac{5}{7}$ are ______________________.

4. Equivalent fraction of $\frac{3}{8}$ with numerator 15 is __________________.

5. Equivalent fraction of $\frac{4}{9}$ with denominator 45 is __________________.

6. Find the missing numerals: $\frac{3}{5} = \frac{\square}{10} = \frac{9}{\square} = \frac{\square}{20} = \frac{15}{\square}$

7. Equivalent fraction of $\frac{10}{15}$ with denominator 3 is _____________.

8. Equivalent fraction of $\frac{16}{28}$ with denominator 7 is ______________.

9. Equivalent fraction of $\frac{21}{27}$ with numerator 7 is ________________.

10. Equivalent fraction of $\frac{15}{55}$ with numerator 3 is ________________.

11. Find the missing numeral: $\frac{9}{15} = \frac{3}{\square}$

12. How many equivalent fractions can be written for a given fraction? __________

Worksheet - 28

Equivalent Fractions, Types of Fractions

1. If cross products of two fractions are equal, they are __________. Also, if ____________ of two fractions are not equal, they are not equivalent.

2. If two fractions are equivalent their cross products are __________. Also, if two fractions are not __________, their cross products are not equal.

3. Check if the following fractions are equivalent. Write Yes or No.

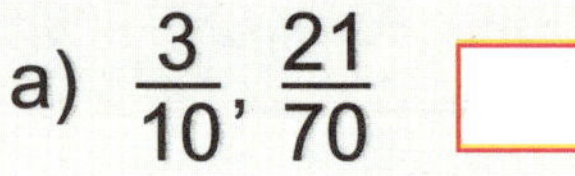

a) $\frac{3}{10}, \frac{21}{70}$ ☐   b) $\frac{5}{9}, \frac{35}{60}$ ☐

4. Fill in the blanks.

a) Fractions like $\frac{1}{2}, \frac{1}{3}$ and $\frac{1}{7}$ are called __________ fractions because they have numerator as 1.

b) Fractions like $\frac{2}{5}, \frac{3}{8}$ and $\frac{4}{9}$ are called ______________ fractions because their numerator is less than their denominator.

c) Fractions like $\frac{4}{3}, \frac{5}{3}$ and $\frac{11}{9}$ are called ______________ fractions because their numerator is more than their denominator.

d) Fractions like $2\frac{1}{2}$, $5\frac{1}{4}$ and $2\frac{1}{3}$ are called ________________ fractions because they have an integer part and a fraction part.

e) Fractions like $\frac{2}{7}, \frac{3}{7}$ and $\frac{4}{7}$ are called _____________ fractions because they have the same denominator.

f) Fractions like $\frac{1}{13}, \frac{5}{7}$ and $\frac{6}{11}$ are called _____________ fractions because they have different denominators.

Worksheet - 29

Reducing to Lowest Terms, Addition and Subtraction of Fractions

1. Fill in the blanks.

a) Reduce $\frac{24}{56}$ to its lowest term ____________________

b) Express $\frac{90}{11}$ as a mixed fraction ____________________

c) Express $4\frac{2}{3}$ as an improper fraction ____________________

2. Add the following:

a) $\frac{2}{7} + \frac{4}{7} =$ ____________________

b) $2\frac{4}{11} + \frac{5}{11} =$ ____________________

c) $2\frac{2}{9} + 1\frac{5}{9} =$ ____________________

3. Subtract the following:

a) $\frac{11}{13} - \frac{9}{13} =$ ____________________

b) $2\frac{7}{13} - \frac{4}{13} =$ ____________________

c) $2\frac{11}{17} - 2\frac{8}{17} =$ ____________________

4. Simplify the following:

a) $7\frac{7}{11} + 2\frac{4}{11} - 4\frac{3}{11} =$ ____________________

b) $9\frac{13}{25} - 2\frac{5}{25} - 3\frac{3}{25} =$ ____________________

c) $6\frac{6}{17} - 2\frac{7}{17} + 3\frac{4}{17} =$ ____________________

Worksheet - 30

Tenths, Hundredths and Thousandths Pictorial Representations

1. Fill in the blanks.

a) 2.985 is read as: ____________________

b) __________ is read as zero point six eight.

2. Express as decimals.

a) $\frac{9}{10}$ = __________ b) $3\frac{9}{100}$ = __________ c) $\frac{67}{1000}$ = __________

3. Write these decimals as fractions or mixed numerals.

a) 2.3 = __________ b) 1.93 = __________ c) 12.473 = __________

4. Write the following numbers in the place value chart:

a) 13.2 b) 7.57 c) 234.079

	Hundreds 100	Tens 10	Ones 1	Tenth $\frac{1}{10}$	Hundredths $\frac{1}{100}$	Thousands $\frac{1}{1000}$
a)						
b)						
c)						

5. Determine the decimal shown in the figure.

a) Decimal = __________

b) 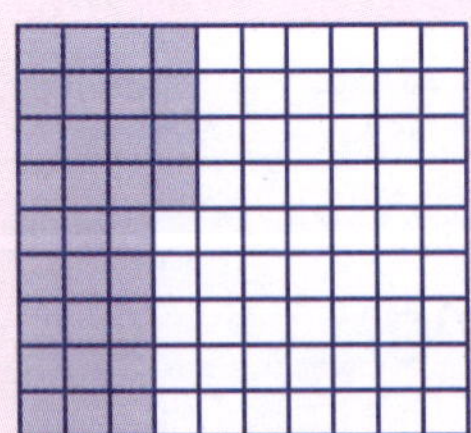 Decimal = __________

6. Represent the decimal in the figure.

a) 0.6 b) 0.89

Unit-5
Metric Measures

Worksheet - 31

Conversion of Metric Units

Fill in the blanks.

1. 28700 cm = ____________ m
2. 3234 m = ____________ cm
3. 45 km = ____________ m
4. 589000 m = ____________ km
5. 630 m 36 cm = ____________ cm
6. 72342 cm = ______ m ______ cm
7. 83 km 322 m = ____________ m
8. 98238 m = ______ km ______ m
9. 12000 g = ____________ kg
10. 26 kg = ____________ g
11. 37852 g = ______ kg ______ g
12. 429800 g = ______ kg ______ g
13. 57 kg 258 g = ____________ g
14. 613 kg 123 g = ____________ g
15. 74000 ml = ____________ L
16. 831000 ml = ____________ L
17. 92 L = ____________ ml
18. 126 L = ____________ ml
19. 22 L 400 ml = ____________ ml
20. 221 L 500 ml = ____________ ml
21. 39750 ml = ______ L ______ ml
22. 113400 ml = ______ L ______ ml

Worksheet - 32

Addition of Metric Units

Solve the following:

1. a)
```
  52 cm
+ 28 cm
_______
     cm
```
b)
```
  74 m
+ 18 m
______
     m
```
c)
```
  825 m  80 cm
+ 116 m  45 cm
______________
      m     cm
```
d)
```
  543 m
+ 336 m
_______
      m
```
e)
```
  257 km
+ 329 km
________
      km
```
f)
```
  362 km  735 m
+ 477 km  982 m
_______________
       km     m
```
g)
```
  326 g
+ 347 g
_______
      g
```
h)
```
  141 kg
+ 756 kg
________
      kg
```
i)
```
  135 kg  250 g
+  46 kg  172 g
_______________
       kg     g
```
j)
```
  450 ml
+ 267 ml
________
      ml
```
k)
```
  2389 L
+ 4178 L
________
       L
```
l)
```
  225 L  230 ml
+ 357 L  290 ml
_______________
      L      ml
```

2. 36300 cm + 24300 cm = ______________ cm or ______________ m

3. 64000 m + 36000 m = ______________ m or ______________ km

4. 27000 g + 52000 g = ______________ g or ______________ kg

5. 84000 g + 32000 g = ______________ g or ______________ kg

6. 46000 ml + 15000 ml = ______________ ml or ______________ L

7. 54000 ml + 32000 ml = ______________ ml or ______________ L

Worksheet - 33

Subtraction of Metric Units

Solve the following:

1. a)
```
   67 cm
 – 45 cm
 -------
   ___ cm
```
b)
```
   99 m
 – 53 m
 ------
   ___ m
```
c)
```
   745 m  67 cm
 – 498 m  12 cm
 --------------
   ___ m  ___ cm
```
d)
```
   743 m
 – 462 m
 -------
   ___ m
```
e)
```
   895 km
 – 699 km
 --------
   ___ km
```
f)
```
   847 km  127 m
 – 239 km  720 m
 ---------------
   ___ km  ___ m
```
g)
```
   726 g
 – 429 g
 -------
   ___ g
```
h)
```
   967 kg
 – 521 kg
 --------
   ___ kg
```
i)
```
   596 kg  281 g
 – 123 kg  810 g
 ---------------
   ___ kg  ___ g
```
j)
```
   589 ml
 – 127 ml
 --------
   ___ ml
```
k)
```
   6720 L
 – 2093 L
 --------
   ___ L
```
l)
```
   856 L 109 ml
 – 127 L 390 ml
 --------------
   ___ L ___ ml
```

2. 16200 cm – 14800 cm = ________________ cm or ________________ m

3. 75000 m – 50000 m = ________________ m or ________________ km

4. 53 kg – 28 kg = ________________ kg or ________________ g

5. 94 kg – 52 kg = ________________ kg or ________________ g

6. 63 L – 25 L = ________________ L or ________________ ml

7. 65 L – 39 L = ________________ L or ________________ ml

Worksheet - 34

Multiplication of Metric Units

Solve the following:

1. a) 24 cm × 5 = ____ cm

b) 118 g × 7 = ____ g

c) 127 ml × 9 = ____ ml

d) 246 m × 8 = ____ m

e) 43 m × 12 = ____ m

f) 335 km × 14 = ____ km

g) 237 kg × 13 = ____ kg

h) 2341 L × 15 = ____ L

i) 13 m 23 cm × 16 = ____ m ____ cm

j) 14 km 123 m × 18 = ____ km ____ m

k) 15 kg 128 g × 17 = ____ kg ____ g

l) 28 L 125 ml × 19 = ____ L ____ ml

2. 1325 cm × 12 = ______________ cm or ______________ m

3. 500 m × 14 = ______________ m or ______________ km

4. 1250 g × 16 = ______________ g or ______________ kg

5. 400 g × 15 = ______________ g or ______________ kg

6. 300 ml × 420 = ______________ ml or ______________ L

7. 700 ml × 20 = ______________ ml or ______________ L

Worksheet - 35

Division of Metric Units

Solve the following:

1. Divide 646 m 68 cm by 12.

646 m 68 cm = __________ cm

_______ cm ÷ 12 = __________ cm

= _______ m _______ cm

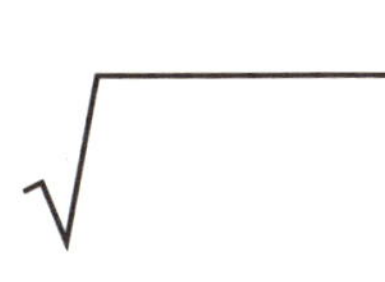

2. Divide 179 km 808 m by 16.

179 km 808 m = __________ m

__________ m ÷ 16 = __________ m

= _______ km _______ m

3. Divide 77 kg 462 g by 14.

77 kg 462 g = __________ g

__________ g ÷ 14 = __________ g

= _______ kg _______ g

4. Divide 568 L 278 ml by 18.

_____ L _____ ml = __________ ml

__________ ml ÷ 18 = _____ ml

= _____ L _____ ml

Worksheet - 36

Word Problems

1. A railway line 125 km long was extended at one end by 12000 m. What is its length in kilometres now?

____

____

2. A bag had 34 kg 700 g of sugar. Of this 5700 grams was used. What is the weight in kilograms of sugar in the bag now?

____

____

3. Each vial is to have 20 ml of medicine. How many litres of medicine are needed to fill 2,000 such vials?

____

____

4. A cable is 800 m long. It is to be cut into 20 pieces of equal length. What will be the length of each piece in centimetres?

____

____

5. A container had 10 kg of chilli powder. It was packed into pouches containing 50 grams each. How many pouches were packed?

____

____

6. A water tank has 200 L of water in it. Out of this 5000 ml water has been used up. What is the remaining quantity of water in the tank in L?

____

____

Worksheet - 37

Estimating Metric Units

1. Round off the following:

a) 123 m 65 cm ≈ ________ m

b) 53 km 465 m ≈ ________ km

c) 23 kg 425 g ≈ ________ kg

d) 86 L 297 ml ≈ ________ L

2. Estimate the following by rounding to indicated unit:

a) 262 m 62 cm + 482 m 79 cm ≈ _____ m + _____ m = _____ m

b) 33 km 251 m – 18 km 742 m ≈ _____ km – _____ km = _____ km

c) 11 kg 364 g + 12 kg 472 g ≈ _____ kg + _____ kg = _____ kg

d) 54 kg 122 g – 28 kg 622 g ≈ _____ kg – _____ kg = _____ kg

e) 31 L 364 ml + 16 L 472 ml ≈ _____ L + _____ L = _____ L

f) 74 L 122 ml – 23 L 622 ml ≈ _____ L – _____ L = _____ L

g) A consignment weighs 18904 kg. Another consignment weighs 15848 kg. Estimate the total weight by rounding to nearest ten kilograms.

____

h) My car's milometer reads 32874 miles. At the start of the year it read 12375 miles. Estimate the distance covered by my car this year by rounding to nearest hundred miles.

____

i) A container has a capacity of 56520 L. Another container has the capacity of 45870 L. Estimate the difference in capacity by rounding to nearest thousand litres.

____

Unit-6
Time and Money

Worksheet - 38

Converting Units of Time, Leap Year, am or pm

1. Fill in the blanks.

a) 6 hours = ____________ minutes
b) 7 days = ____________ hours
c) 5 weeks = ____________ days
d) 8 years = ____________ months
e) 2 decades = ____________ years
f) Tick the leap years out of the following:

2008 ☐ 2014 ☐ 2016 ☐ 2026 ☐ 2032 ☐

g) The year 2004 was a leap year. The next leap year will be ____________.
h) The year 2024 will be a leap year. The leap year just before it will be ____________.

2. Write the time using am or pm.

a) Half past 5 in the morning ____________
b) Quarter to 6 in the evening ____________
c) Quarter past 8 in the morning ____________
d) Quarter to 8 in the evening ____________
e) 2 hours before 12 noon ____________
f) 1 hour after 12 midnight ____________
g) 11:30 in morning ____________
h) 7:30 in evening ____________

3. Write the time using morning or afternoon or evening.

a) 4:45 pm means quarter to ____________ in the ____________.
b) 2:30 pm means half past ____________ in the ____________.
c) 6:30 am means half past ____________ in the ____________.

Worksheet - 39

Time on Clocks, Days in Between

1. Write the time shown on each clock.

a	b	c

______:______	______:______	______:______

2. Draw hands on the clock to show the indicated time.

a	b	c
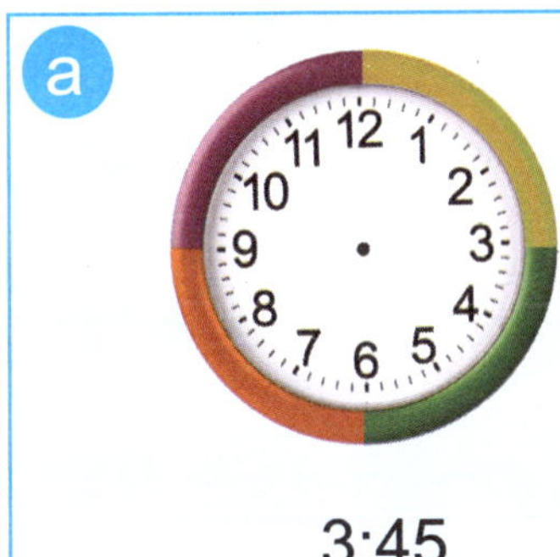		
3:45	6:40	7:05

3. Fill in the blanks.

a) Time shown by clock if hour hand is at 9 and minute hand is at 2 is ______.

b) If a clock shows 2:35, hour hand is between ________ and ________ and minute hand is at ________.

c) A clock shows 11:55. The time is nearly ________ o' clock.

d) A clock shows 8:05. Time correct to nearest hour is ___________.

e) A clock shows 6:25. The time is nearly half past ___________.

f) A clock shows 7:40. The time is nearly quarter to ___________.

g) A clock shows 8:20. The time is nearly quarter past ___________.

h) There are ___________ days in between March 15 and April 15.

i) There are 12 days in between March 10 and ______________.

j) There are 13 days in between ___________ and September 22.

Worksheet - 40

Conversion (Dollar and Cents), Operations on Money

1. Fill in the blanks.

a) \$ 321 = _______ ¢

b) \$ 166.32 = _______ ¢

c) 36000 ¢ = \$ _______

d) 17400 ¢ = \$ _______

2. Solve the following:

a)
```
  $44733
 +$22567
 _______
  $
 _______
```

b)
```
  $1737.26
 +$2948.41
 _________
  $
 _________
```

c)
```
  $75201
 -$25293
 _______
  $
 _______
```

d)
```
  $4425.66
 -$2134.07
 _________
  $
 _________
```

e)
```
  $7421
    ×12
 ______

 ______
  $
 ______
```

f)
```
  $122.41
      ×19
 ________

 ________
  $
 ________
```

g) \$ 37264 ÷ 17

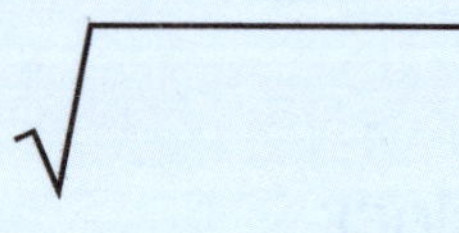

\$ _______________

h) \$ 1951.04 ÷ 14

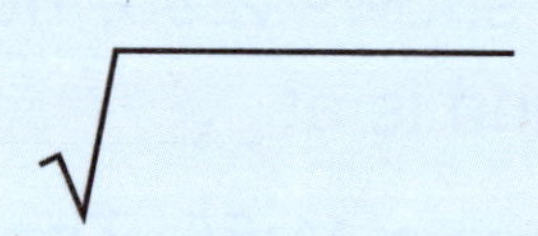

\$ _______________

Worksheet - 41

Word Problems

1. Maria had ₹ 210,500 in her bank account. She deposited ₹ 13,500 in her bank account. What is the amount in her account now?

___

___

2. A scooter costs ₹ 42,450 and a motorcycle ₹ 62,735. How much more does the motorcycle cost than the scooter?

___

___

3. A company makes handkerchiefs. Cost of making one piece is ₹ 15. How much will it cost to make 5,300 handkerchiefs?

___

___

4. The cost of 18 washing machines is ₹ 226,080. What is the cost of one washing machine?

___

___

5. Jack had ₹ 234,500. He made a payment of ₹ 45,000. Then he received a payment of ₹ 65,500. What amount does he have with him now?

___

___

6. Lara had ₹ 184,000 with her. She bought a purse for ₹ 6,400 and a dress for ₹ 4,800. What is the balance amount with her now?

___

___

Worksheet - 42

Estimating Money

1. **Round the following to the nearest dollar:**

   a) $ 422.86 ≈ $ __________ b) $ 1224.36 ≈ $ __________

2. **Estimate the following by rounding to nearest dollar:**

   a) $ 551.45 + $ 363.84 ≈ $ __________ + $ __________ = $ __________

   b) $ 249.86 – $ 232.73 ≈ $ __________ – $ __________ = $ __________

3. **Estimate the following by rounding to nearest ten dollars:**

   a) $ 78121 + $ 12217 ≈ $ __________ + $ __________ = $ __________

   b) $ 32125 – $ 17335 ≈ $ __________ – $ __________ = $ __________

4. **Estimate the following by rounding to nearest hundred dollars:**

   a) $ 28374 + $ 64731 ≈ $ __________ + $ __________ = $ __________

   b) $ 69744 – $ 36421 ≈ $ __________ – $ __________ = $ __________

5. **Estimate the following by rounding to nearest thousand dollars:**

   a) $ 36444 + $ 52249 ≈ $ __________ + $ __________ = $ __________

   b) $ 53398 – $ 14842 ≈ $ __________ – $ __________ = $ __________

6. **Estimate the following by rounding to nearest ten thousand dollars:**

   a) $ 36449 + $ 55238 ≈ $ __________ + $ __________ = $ __________

   b) $ 734789 – $ 154824 ≈ $ __________ – $ __________ = $ __________

7. **Estimate the following by rounding to nearest hundred thousand dollars:**

   a) $ 298300 + $ 420098 ≈ $ __________ + $ __________ = $ __________

   b) $ 786500 – $ 239800 ≈ $ __________ – $ __________ = $ __________

Point, Line Segment, Line and Ray

Fill in the blanks.

1. An impression by the tip of a sharp pencil represents a ______________.
2. A line segment has a definite _________ and _________ end points.
3. Line segment AB can also be named as line segment ______________.
4. If a line segment is stretched only on one side we get a ______________.
5. Ray AB is written as ______________.
6. The starting point of the ray is also called its ______________ point.
7. We cannot find the ________ of ray AB as it extends endlessly in one direction.
8. Ray AB and ray BA are in ______________ directions.
9. If a line segment is stretched on both sides we get a ______________.
10. Line AB is written as ______________.
11. A line has no ______________ points.
12. We cannot find the ________ of line AB as it extends endlessly on both sides.
13. Line AB can also be named as line ______________.
14. An infinite number of __________ can be drawn to pass through a fixed point.
15. Through ______________ given points only one line can be drawn.
16. A, B and C are three different points. Taking two points at a time we can draw at least ______________ line(s) through them.
17. A, B and C are three different points. Taking two at a time we can draw at most ______________ line(s) through them.
18. A, B, C and D are four points such that ABCD is a quadrilateral. Taking two points at a time we can draw ______________ line segments.
19. Corners of a triangle represent ______________.
20. Edges of a cube represent ______________.

Worksheet - 44

Angles

Fill in the blanks.

1. Two different rays with same __________ form an angle.
2. The two __________ forming an angle are called the arms of the angle.
3. The common end point of the two rays (arms) is called the __________ of the angle.

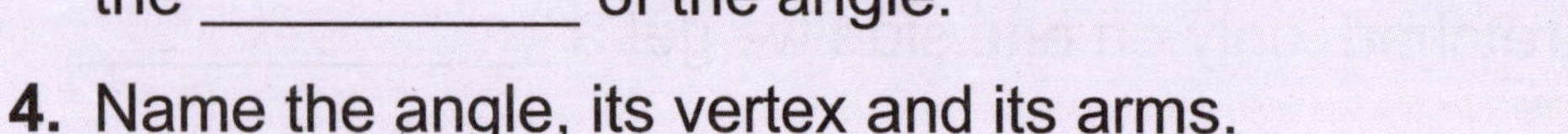

4. Name the angle, its vertex and its arms.

Arms: ____________ Vertex: ____________

Name of angle: ____________

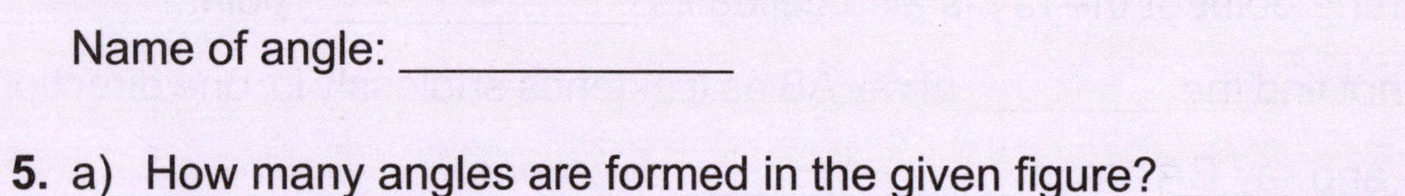

5. a) How many angles are formed in the given figure? __________

   b) Name each angle. ____________________

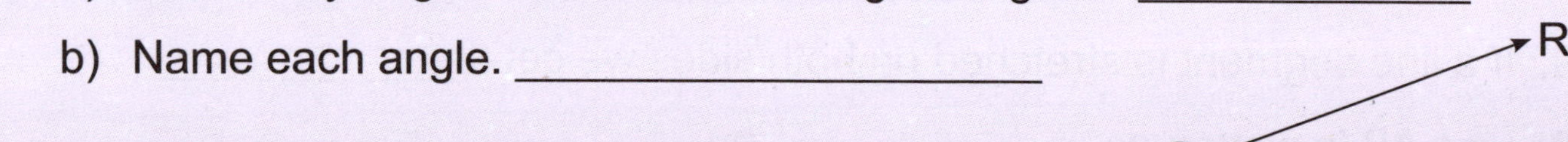

____

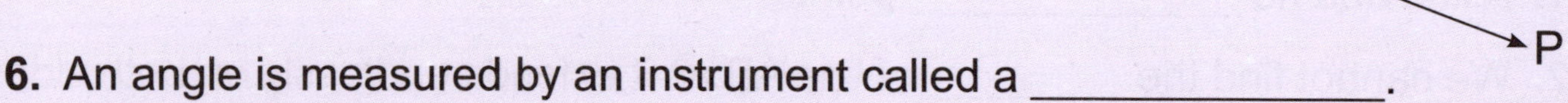

6. An angle is measured by an instrument called a ____________.
7. The instrument's circular edge has a ________ scale and an ________ scale.
8. The inner scale is marked from 0 to 180 in ____________ direction.
9. The outer scale is marked form 0 to 180 in ____________ direction.
10. Each mark represents __________ degree and is written as __________.
11. Measure each of the following angles with a protractor:

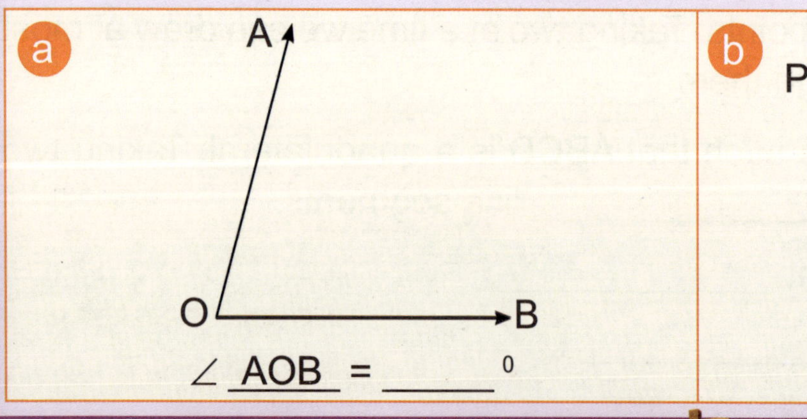

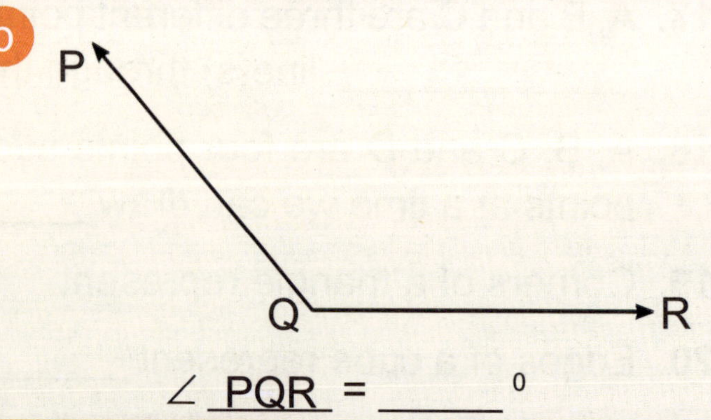

Worksheet - 45

Circle, Semicircle, Symmetry, Nets

1. Fill in the blanks.

a) Midpoint of a diameter of a circle is the ____________ of a circle.

b) Diameter = 2 × ____________

c) AB is the diameter of a circle and O is its midpoint. If AB = 8 cm then OB = ____________ cm

d) A diameter of a circle divides it into ____________ semicircles.

e) Draw the line of symmetry for each of the following:

A	T	X	K

2. Given below is the net of a cube. What is the side of the cube?

Side of cube = _____ cm

3. Given below is the net of a cuboid. What are its length, breadth and height?

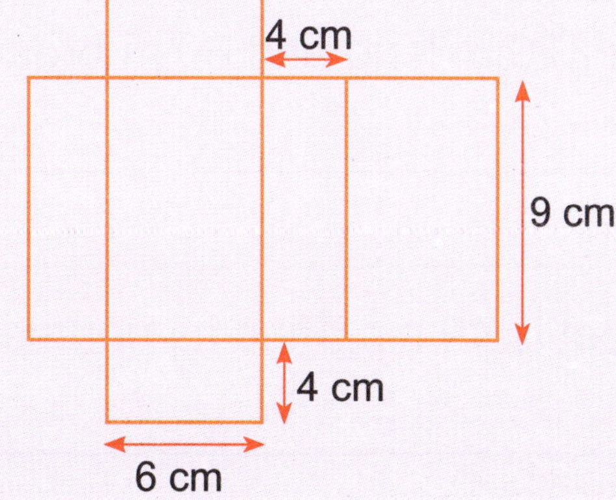

Length = _________ cm, Breadth = _________ cm, Height = _________ cm

Worksheet - 46

Perimeter of Triangle, Quadrilateral, Rectangle and Square

Fill in the blanks.

1. Perimeter of a triangle with sides 6 cm, 5 cm and 8 cm is ____________ cm.
2. Perimeter of a triangle is 18 cm. Two of its sides are 8 cm and 4 cm. The length of the third side is ____________ cm.
3. Perimeter of a triangle is 23 m. One of its side is 11 m and the other two sides are equal. The length of each equal side is ____________ m.
4. Perimeter of a triangle is 33 cm. If all three sides are equal then length of each side is ____________ cm.
5. Perimeter of a quadrilateral with sides 8 cm, 12 cm, 14 cm and 10 cm is ____________ cm.
6. Perimeter of a quadrilateral is 43 cm. Three of its sides are 10 cm, 9 cm and 12 cm. Length of the remaining side is ____________ cm.
7. Perimeter of a quadrilateral is 28 m. One of its side is 10 m and the other three sides are equal. The length of each equal side is ____________ m.
8. Perimeter of a quadrilateral is 48 cm. If all four sides are equal then length of each side is ____________ cm.
9. A rectangle is 10 cm long and 6 cm broad. Find its perimeter and area.

   ____

   ____

10. A rectangular park is 120 m long and 60 m broad. Find its perimeter and area.

    ____

    ____

11. Side of a square sheet is 49 cm. Find its perimeter and area.

    ____

    ____

Worksheet - 47

Perimeter of Polygon

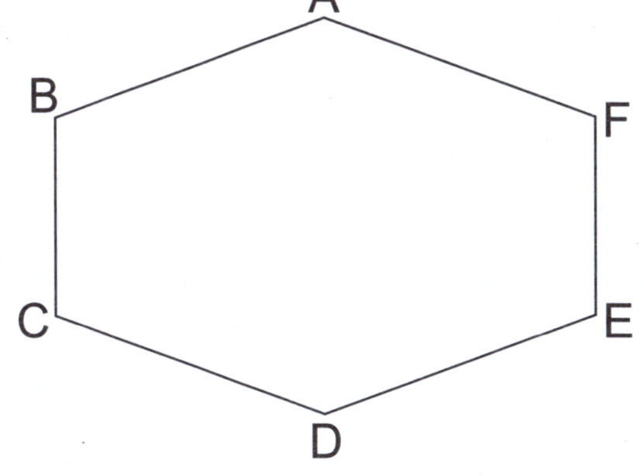

1. Perimeter of polygon ABCDEF

= ______ + ______ + ______

\+ ______ + ______ + ______

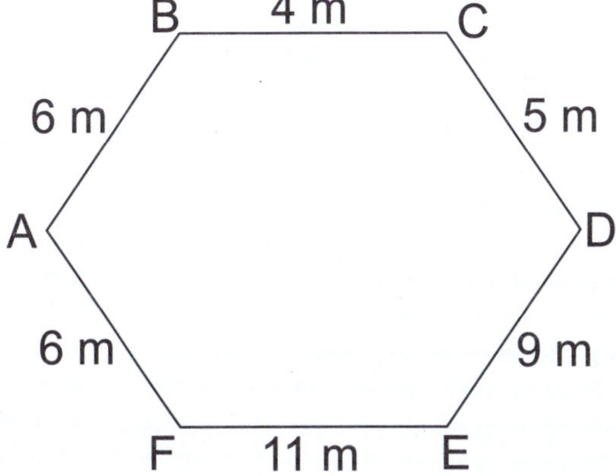

2. Find the perimeter of polygon ABCDEF.

____

____

____

____

3. A polygon has five sides. Four of the sides are 10 m, 9 m, 8 m and 7 m and perimeter is 40 m. Find the fifth side of the polygon.

____

____

4. A field is in the shape of a polygon with five equal sides. It took 200 m of wire to fence it from all sides. What is the length of each side of the field?

____

____

5. A polygon has 6 sides. Its perimeter is 2 m. Four of its sides are 20 cm, 25 cm, 35 cm and 40 cm. Remaining sides are equal. Find the length of each of the equal side.

____

____

Worksheet - 48

Reading and Drawing Bar Charts

1. **The bar graph given below shows the marks obtained by students in a class test of 10 marks.**

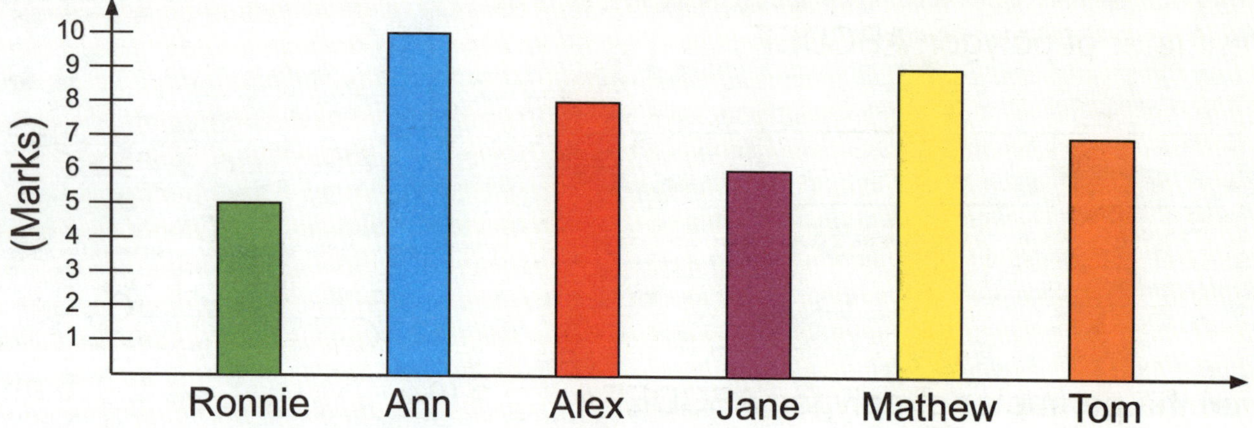

Using the above graph answer the following:

a) Who scored least marks? ______

b) Who scored 7 marks? ______

c) How many marks less did Ronnie score than Jane? ______

d) Who scored the highest marks? ______

e) How many students scored less than 7 marks? ______

f) Who scored more than Alex but less than Ann? ______

2. **In an election, parties A, B, C, D and E won the following number of seats:**

Party	A	B	C	D	E
Number of seats	2	12	7	8	5

Represent the above data as a bar chart (draw horizontal bars).

parties

number of seats

Worksheet - 49

Complete the Patterns

1. Draw the next three figures for each of the following:

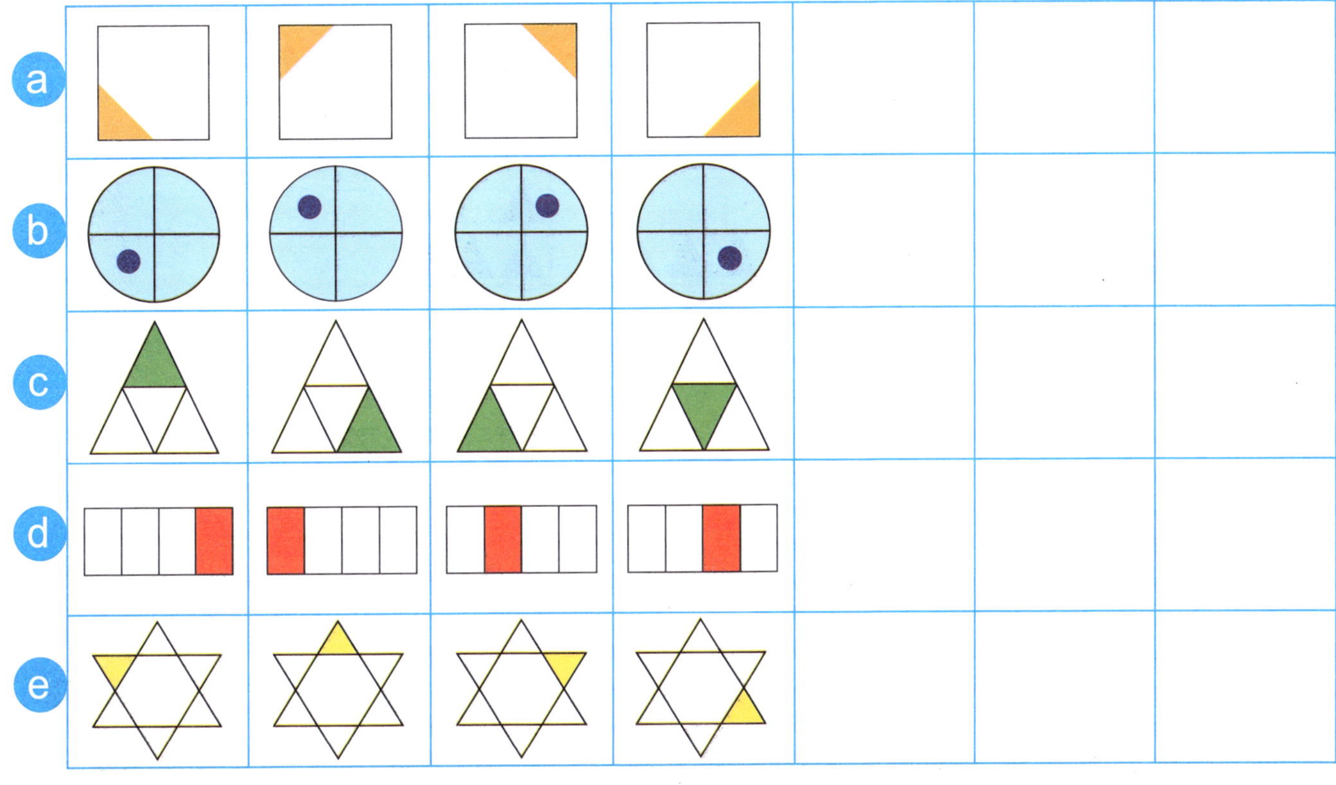

2. Fill in the blank boxes to complete the pattern.

Worksheet - 50

Growing and Reducing Patterns

1. Draw the next two figures for each of the following:

a

b

c

d

e

f

2. Extend the sequence up to two terms.

a B9*b*, D8*d*, F7*f*, H6*h*,	
b 1122, 2233, 3344, 4455,	
c 1230, 2340, 3450, 4560,	
d A1Z, B2Y, C3X, D4W,	
e PQRS, QRST, RSTU, STUV,	
f AABB, BBCC, CCDD, DDEE,	

Unit-1 Numbers (Above 10000)

Worksheet-1

1. 2 ten thousands 7 thousands 6 hundreds 5 tens 3 ones
2. 2 hundred thousands 3 ten thousands 8 thousands 9 hundreds 2 tens 5 ones
3. a) 65 435 b) 835 328
4. a) PV of 9 is 90 000, PV of 5 is 5 000, PV of 1 is 100, PV of 2 is 20, PV of 7 is 7
   b) PV of 3 is 300 000, PV of 4 is 40 000, PV of 5 is 5 000, PV of 7 is 700, PV of 8 is 80, PV of 9 is 9
5.

	H-Th	T-Th	Th	H	T	O
a)		6	9	7	5	8
b)	7	5	3	5	4	6

6. a) Twelve thousand nine hundred eighty-three
   b) Two hundred thirty seven thousand four hundred fifty nine

Worksheet-2

1. a) 63 807 b) 150 220
2. a) 70000 + 1000 + 700 + 30 + 4
   b) 300000 + 40000 + 5000 + 700 + 8
3. a) 73478 b) 456048
4. a) 87315 b) 652121 c) 52440, 52441
   d) 759310, 759311, 759312
5. a) 75427 b) 552846 c) 39801, 39800
   d) 478922, 478921, 478920
6. a) 24183, 24184
   b) 630318, 630319, 630320
   c) 235193, 235194, 235195, 235196

Worksheet-3

1. 25300, 25500, 25700
2. 228000, 233000, 238000
3. 40001, 50001, 60001
4. 900000, 1200000, 1500000
5. 241521
6. 43134
7. 442727, 443257, 443329, 457420
8. 44999, 44845, 34777, 32343
9. 13468, 86431
10. 12579, 97521
11. 20000, 99999
12. 10346, 76431
13. 222222, 888888

Worksheet-4

1. 102357, 753201
2. 111111, 888888
3. 102456, 765421
4. a) 98789 b) 700001 c) 785782
   d) 12453 e) 12220 f) 864210
   g) 12345 h) 987654 i) 33333
5. a) 37430 b) 129030
6. a) 47700 b) 439000
7. a) 56000 b) 871000
8. a) 60000 b) 980000
9. a) 700000 b) 900000

Unit-2 Operations on Numbers

Worksheet-5

1. 67877 2. 897967 3. 87799 4. 988999
5. 13259 6. 111550 7. 79057 8. 921214
9. 651003 10. 122078

Worksheet-6

1. 384,233 2. 113,971 3. 290,346
4. 266,603 5. 126,637 6. 90,295

Worksheet-7

1. 24324
2. 56646; 56562
3. 23074
4. 74214; 83985
5. 42546; 54284; 37412
6. 62352 7. 0 8. 46679
9. 74757 10. 0 11. 33169
12.

a

	5	3	5	2	1
+	3	1	4	4	6
	8	4	9	6	7

b

	6	5	3	2	3	5
+	3	4	4	3	2	4
	9	9	7	5	5	9

c

	1	1	1		
	5	7	8	9	5
+	3	4	6	9	2
	9	2	5	8	7

d

	1	1	1	1	1	
	3	2	3	6	7	2
+	2	9	7	8	7	9
	6	2	1	5	5	1

Worksheet-8

1. 73340 2. 34,840 3. 97700 4. 26,200
5. 341000 6. 77,000 7. 230000 8. 370,000
9. 900000 10. 700,000

Worksheet-9

1. 34131 2. 62244 3. 303221 4. 562227
5. 7898 6. 26174 7. 507436 8. 735839
9. 74091 10. 29446

Worksheet-10

1. 3,124 2. 80,880 3. 79,930
4. 98,384 5. 299,679 6. 102,142

Worksheet-11

1. a) 22838 b) 8,715
2. a) 0 b) 562376 c) 49214
   d) 288352 e) 0 f) 867106
3. correct 4. correct
5. a)

	7	8	9	5	5
–	4	5	7	3	1
	3	3	2	2	4

b)

	4	5	4	8	3	3
–	3	1	5	4	7	6
	1	3	9	3	5	7

Worksheet-12

1. 40000 2. 10,000 3. 300000 4. 300,000
5. 17700 6. 2,500 7. 337000 8. 218,000
9. 19670 10. 170,530

Worksheet-13

1.

1	11	12	13	14	15	16	17	18	19	20
2	22	24	26	28	30	32	34	36	38	40
3	33	36	39	42	45	48	51	54	57	60
4	44	48	52	56	60	64	68	72	76	80
5	55	60	65	70	75	80	85	90	95	100
6	66	72	78	84	90	96	102	108	114	120
7	77	84	91	98	105	112	119	126	133	140
8	88	96	104	112	120	128	136	144	152	160
9	99	108	117	126	135	144	153	162	171	180
10	110	120	130	140	150	160	170	180	190	200

2. 128 3. 119 4. 162 5. 190 6. 120

Worksheet-14

1. 63679 2. 77544 3. 71903
4. 69818 5. 686535 6. 879568
7. 609331 8. 815706 9. 681017

Worksheet-15

1. 89650 2. 309624 3. 693611
4. 116424 5. 364014 6. 172569
7. 977664 8. 376428 9. 604821

Worksheet-16

1. 232,000 2. 28,224 3. 86,950
4. 332,100 5. 55,350 6. 59,400

Worksheet-17

1. a) 13490 b) 271500 c) 49000
   d) 520000 e) 500000 f) 900000
   g) 34,250 h) 452,300 i) 113,000
   j) 140,000 k) 800,000
2. a) 24210 b) 309000 c) 54000
   d) 390000 e) 50,200 f) 63,900
   g) 46,000 h) 80,000

Worksheet-18

1. a) 43200 b) 63000 c) 60000
   d) 80000 e) 12,600 f) 95,000
   g) 960,000 h) 800,000 i) 28563
   j) 14349 k) 50241; 12874
   l) 20024; 57465; 42140 m) 1
   n) 47557 o) 0 p) 0
2. a) 29400 b) 218000 c) 13,500
3. a) 10000 b) 40,000

Worksheet-19

1. 10872 2. 5392 3. 46085 4. 14328

Worksheet-20

1. 9569; 6 2. 6329; 3 3. 5486; 38 4. 964; 13

Worksheet-21

1. 3,109 2. 1,359 3. 1,068 4. 1,826

Worksheet-22

1. 3,128; 7 2. 4,209; 6 3. 7,818; 10 4. 419; 125

Worksheet-23

1. a) 1587 b) 69764 c) 249 d) 7238 e) 32
   f) 812 g) 4 h) 99 i) 5 j) 80
2. a) 6 b) 8
3. a) 1 b) 133413 c) 52545 d) 624511
   e) 1 f) 157437 g) 0 h) 0
   i) 62730; 3690; 17 j) 11012; 21; 524; 8

Unit-3 Multiples and Factors

Worksheet-24

1. a) 18, 36, 54, 72, 90 b) 80, 96, 112, 128, 144
   c) 60 and 80 d) 96 e) 99 f) 114
   g) 140 h) 14348, 14350
   i) 15435, 15437 j) 12366, 12368
   k) 52399, 52401 l) 98 m) 101 n) 1000
   o) 99998 p) 999999
2. a) True b) False c) True d) True
3. Even: 1234, 27872, 48418, 5686, 901294
   Odd: 5681, 3403, 7187, 6729, 85955

Worksheet-25

1. a) 10234 b) 987653 c) 102346
2. a) 1, 2, 5 and 10 b) 1, 2, 4, 8, 16 and 32
   c) 1, 2, 3, 4, 6, 9, 12, 18 and 36. d) 1, 23
3. Yes

4. a) 1, 3, 5 and 15 b) 1, 37 c) 1, 7 d) 1, 5
5. Yes
6. a) 2, 4, 22 and 44 b) 2, 10 and 50
   c) No even factor d) No even factor
7. No
8. a) 12, 16, 18, 24, 36, 48 and 72
   b) 10, 20, 25, 40 and 50
9. Yes; 100 is a factor of 200

Unit-4 Fractions and Decimals

Worksheet-26

1. a) 2/3 b) 3/4
2. a) two-fifth b) three-eighth
3. a) 12; 19 b) 3/7 c) 3/10 d) 4/13
   e) 2/7 f) 8/11 g) smaller h) greater
4. a) < b) > c) <
5. 1/8, 5/8, 7/8 6. 5/7, 5/8, 5/11

Worksheet-27

1. dividing, multiplying 2. 8/28, 10/35 and 12/42
3. 10/14, 15/21 and 20/28 4. 15/40
5. 20/45 6. 6/10 = 9/15 = 12/20 = 15/25
7. 2/3 8. 4/7 9. 7/9 10. 3/11
11. 5 12. Uncountable

Worksheet-28

1. equivalent, cross products
2. equal, equal 3. a) Yes b) No
4. a) unit b) proper c) improper d) mixed
   e) like f) unlike

Worksheet-29

1. a) 3/7 b) $8\frac{2}{11}$ c) 14/3
2. a) 6/7 b) $2\frac{9}{11}$ c) $3\frac{7}{9}$
3. a) 2/13 b) $2\frac{3}{13}$ c) $\frac{3}{17}$
4. a) $5\frac{8}{11}$ b) $4\frac{1}{5}$ c) $7\frac{3}{17}$

Worksheet-30

1. a) Two point nine eight five b) 0.68
2. a) 0.9 b) 3.09 c) 0.067
3. a) $2\frac{3}{10}$ b) $1\frac{93}{100}$ c) $12\frac{473}{1000}$
4.

	Hundreds 100	Tens 10	Ones 1	Tenth $\frac{1}{10}$	Hundredths $\frac{1}{100}$	Thousands $\frac{1}{1000}$
a)		1	3	2		
b)			7	5	7	
c)	2	3	4	0	7	9

5. a) 0.7 b) 0.34

6. a) b)

Unit-5 Metric Measures

Worksheet-31

1. 287 2. 323400 3. 45000 4. 589
5. 63036 6. 723; 42 7. 83322 8. 98; 238
9. 12 10. 26000 11. 37; 852 12. 429; 800
13. 57258 14. 613123 15. 74 16. 831
17. 92000 18. 126000 19. 22400 20. 221500
21. 39; 750 22. 113; 400

Worksheet-32

1. a) 80 b) 92 c) 942, 25 d) 879
   e) 586 f) 840, 717 g) 673 h) 897
   i) 181, 422 j) 717 k) 6567 l) 582, 520
2. 60600; 606 3. 100000; 100 4. 79000; 79
5. 116000; 116 6. 61000; 61 7. 86000; 86

Worksheet-33

1. a) 22 b) 46 c) 247, 55 d) 281
   e) 196 f) 607, 407 g) 297 h) 446
   i) 472, 471 j) 462 k) 4627 l) 728, 719
2. 1400; 14 3. 25000; 25 4. 25; 25000
5. 42; 42000 6. 38; 38000 7. 26; 26000

Worksheet-34

1. a) 120 b) 826 c) 1143 d) 1968 e) 516
   f) 4690 g) 3081 h) 35115 i) 211m 68cm
   j) 254km 214m k) 257kg 176g l) 534L 375ml
2. 15900; 159 3. 7000; 7 4. 20000; 20
5. 6000; 6 6. 126000; 126 7. 14000; 14

Worksheet-35

1. 53, 89 2. 11, 238 3. 5, 533 4. 31, 571

Worksheet-36

1. 137 km 2. 29 kg 3. 40 L
4. 4000 cm 5. 200 6. 195 L

Worksheet-37

1. a) 124 b) 53 c) 23 d) 86
2. a) 746 b) 14 c) 23 d) 25 e) 47 f) 50
   g) 34750 kg h) 20500 miles i) 11000 L

Unit-6 Time and Money

Worksheet-38

1. a) 360 b) 168 c) 35 d) 96
   e) 20 f) 2008; 2016; 2032 g) 2008
   h) 2020

2. a) 5:30 am b) 5:45 pm c) 8:15 am
d) 7:45 pm e) 10:00 am f) 1:00 am
g) 11:30 am h) 7:30 pm
3. a) 5; evening b) 2; afternoon c) 6; morning

Worksheet-39

1. a) 5:40 b) 10:35 c) 7:10
2. a) b) c)
3. a) 9:10 b) 2; 3; 7 c) 12 d) 8
e) 6 f) 8 g) 8 h) 30
i) March 23 j) September 08

Worksheet-40

1. a) 32100 b) 16632 c) 360 d) 174
2. a) 67300 b) 4685.67 c) 49908 d) 2291.59
e) 89052 f) 2325.79 g) 2192 h) 139.36

Worksheet-41

1. ₹ 224,000 2. ₹ 20,285 3. ₹ 79,500
4. ₹ 12,560 5. ₹ 255,000 6. ₹ 172,800

Worksheet-42

1. a) 423 b) 1224 2. a) 915 b) 17
3. a) 90340 b) 14790 4. a) 93100 b) 33300
5. a) 88000 b) 38000 6. a) 100000 b) 580000
7. a) 700000 b) 600000

Unit-7 Geometry

Worksheet-43

1. point 2. length, two 3. BA 4. ray
5. $\overrightarrow{AB}$ 6. initial 7. length 8. opposite
9. line 10. $\overleftrightarrow{AB}$ 11. end 12. length
13. BA 14. lines 15. two 16. one
17. three 18. six 19. points
20. line segments

Worksheet-44

1. initial point 2. rays 3. vertex
4. QP and QR; Q; ∠PQR or ∠RQP
5. a) 3 b) ∠ROQ, ∠QOP, ∠ROP 6. protractor
7. clockwise; anticlockwise 8. anticlockwise
9. clockwise 10. one; 1° 11. a) 75° b) 130°

Worksheet-45

1. a) centre b) radius c) 4 d) two
e) diameter f) A T X K
2. 4 3. 9; 6; 4

Worksheet-46

1. 19 2. 6 3. 6 4. 11
5. 44 6. 12 7. 6 8. 12
9. 32 cm; 60 sq. cm 10. 360 m; 7200 sq.m
11. 196 cm; 2401 sq.cm

Worksheet-47

1. AB + BC + CD + DE + EF + FA 2. 41 m
3. 6 m 4. 40 m 5. 40 cm, 40 cm

Unit-8 Data Handling and Patterns

Worksheet-48

1. a) Ronnie b) Tom c) 1 mark
d) Ann e) 2 f) Mathew
2.

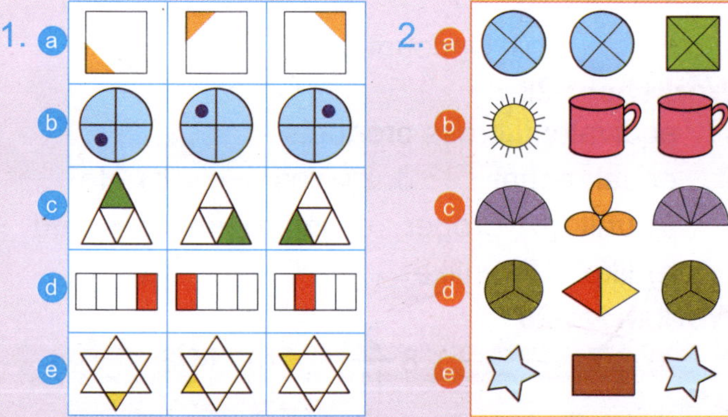

Worksheet-49

1. a b c d e
2. a b c d e

Worksheet-50

1.

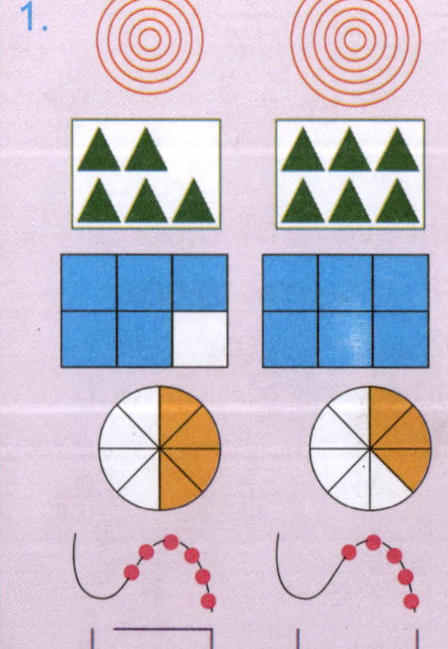

2. a) J5*j*, L4*l*
b) 5566, 6677
c) 5670, 6780
d) E5V, F6U
e) TUVW, UVWX
f) EEFF, FFGG